ELECTRICITY

ROBERT SNEDDEN

Wayland

CONTENTS

4   Introduction

6   Elektron and lodestone

8   Electric machines

10   Down to Earth

12   Galvani and Volta

14   Living electricity

16   Electrochemistry

18   Electricity and magnetism

First published in 1994 by
Wayland (Publishers) Limited
61 Western Road, Hove, East
Sussex BN3 1JD, England
© Copyright 1994 Wayland
(Publishers) Limited

British Library Cataloguing in
Publication Data.
Snedden, Robert
Electricity – (Science Discovery
Series)
I. Title II. Series
537

ISBN 0-7502-1234-9

Acknowledgements
Concept David Jefferis
Text editor and picture research
Michael Brown
Illustrations Robert Burns and
James Robins

Picture credits
Bruce Coleman Ltd 15 (B)
Delta Archive 6, 8, 9(TL), 12,
15(TL), 16(TL), 20, 23, 24,
25(TR), 26, 30, 31, 32, 35,
36(B), 37, 47
Mary Evans Picture Library 16
(BL)
Science Photo Library 4, 9(TR),
11, 13, 14, 19, 25, 27, 28, 29,
39, 40, 41, 45 and all cover
photos
The Science Museum 36(TR)

Printed and bound in Italy by
G. Canale & C.S.p.A., Turin

537x2

NORTHAMPTONSHIRE LIBRARIES	
BFS	0 4 OCT 1996
C537	£9.99

20	Telegraph and relay	36	The electronic age
22	The electric motor	38	Electricity from light
24	The generator	40	New frontiers
26	Lighting up the world	42	Timeline of advance
28	Electricity for all	44	Glossary/1
30	Portable power	46	Glossary/2
32	Electromagnetic waves	47	Going further
34	The cathode-ray tube	48	Index

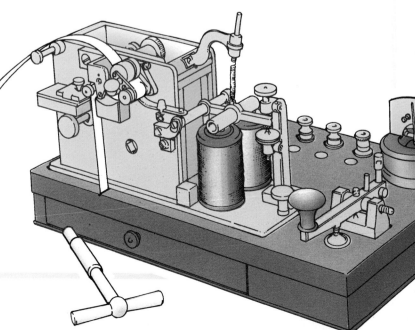

INTRODUCTION

Electricity is easy to take for granted. You press a button or a switch and the television comes on, or a light shines, or the oven heats up. Most of us flick that switch without stopping to think about what happens when we do so, but how does electricity work? How is it made? What exactly is it?

Some of the effects of the linked forces of electricity and magnetism have been known for thousands of years. However, it was a long time before people began to examine them in a truly scientific manner. In this book we follow the steps of the scientists who have pieced together the puzzle of electricity. We will also see how the practical benefits that electricity gives us came about and look at some of the inventions that made it possible.

Even today there are still new developments, and we finish off the book with a look at some possibilities for the future.

▶ The power of electricity allows colourful neon lighting in city centres worldwide.

WHAT IS ELECTRICITY?

Learning about electricity means knowing a little about the structure of matter itself. Everything in the universe – the stars, the planets, this book, your body – is made up of invisibly small atoms.

Each atom has a central nucleus, surrounded by even smaller particles called electrons, orbiting the nucleus. A material such as a metal, that passes electricity easily, is called a conductor and has 'free' electrons moving outside the others. When a metal wire is connected to a battery, the free electrons start to drift along the wire, moving from atom to atom. This movement of electrons is called an electric current.

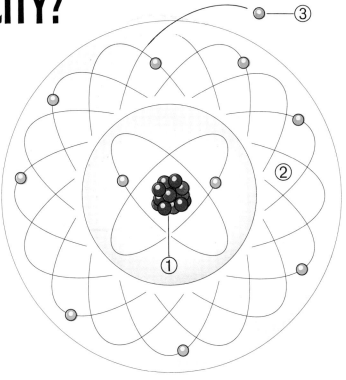

▲ The atom, with its central nucleus (1), surrounded by electrons (2) and a free electron (3).

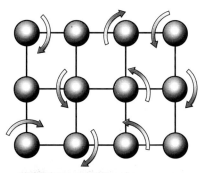

◄ Atoms form a strong, regular pattern in metals. Normally, free electrons move from atom to atom at random.

► When a metal wire is joined to a power source such as a battery, free electrons drift in one direction, as a current. There has to be a complete path to and from the battery. This is called a circuit, and if it is broken, the current stops immediately.

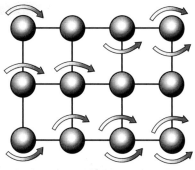

The study of current electricity and magnetic fields is called electrodynamics. There is another type of electricity, where there is no regular flow as such. Instead, electrons build up in one place, hence the name static electricity. Examples of static electricity include thunderstorms and other natural phenomena, where friction transfers electrons from one object to another. The study of this is called electrostatics.

It is impossible to imagine how tiny atoms and electrons are – to take an example, the current flowing through the glowing filament of an ordinary light bulb consists of about a billion billion electrons a second!

This knowledge, simple as it seems, has taken a long time to piece together. The story begins back in the days of ancient Greece...

ELEKTRON AND LODESTONE

The people of ancient Greece knew of a yellow resin produced by some types of tree. They called this resin *elektron*, a substance we know today as amber. When amber sets hard it can be used for jewellery and ornaments. The Greeks, among them the great scientist Thales of Miletus (624-546 BC), found that when amber was rubbed by fur or a piece of cloth it acquired a strange ability. Hair, feathers and other small pieces of material were attracted to the amber and stuck to it, by a force that we know today as static electricity. Thales is often considered to be the founder of science. He was the first recorded person to try to make sense of the world around him without resorting to myth and legend for explanations. He was a great astronomer and could calculate when an eclipse of the Sun would take place.

▶ You can create static electricity by rubbing a balloon with a dry cloth. You will find that you can pick up bits of paper.

▲ Iron filings gather round the two ends, or poles, of a bar magnet. In doing so, they reveal the shape of its magnetic field.

Thales was also aware of the properties of the lodestone, a mineral containing iron, which is a strong natural magnet. Thales was the first person we know of who studied magnetism, but as far as we know he didn't make a connection between the attractive properties of amber and lodestone. Other early thinkers on electricity and magnetism included the Roman philosopher Lucretius (100-55 BC), who spotted that iron filings gathered in a regular pattern near a magnet, but it was not until the sixteenth century that serious studies began.

▶ William Gilbert

Even if it was not understood why lodestone worked, there was a good use for it. If a lodestone shaped like a thin bar or a needle was allowed to move freely, either floating in water on a wood float or suspended from a cord, it always pointed in a roughly north-south direction. It could be used to find which way you were going, and this was the discovery that led to the invention of the compass. Vikings used lodestones on their raiding and exploring voyages, and by AD 1200 the compass was a regular navigation aid for sailors.

William Gilbert (1544-1603), English court physician to Queen Elizabeth I, wrote a book called *De Magnete*, which means 'About Magnets'. Published in 1600, it put forward the theory that the Earth was a giant lodestone, with north and south magnetic poles. This book collected together everything that was known at that time about electricity and magnetism. Gilbert, in fact, was the first person to use the name electricity, from the Greek elektron. He suggested that electricity and magnetism were actually aspects of the same force. He also pointed out the difference between amber and lodestone – amber could attract any light material, but lodestone could only attract iron. Although Gilbert couldn't explain the things he described, his work provided a real starting point for the study of electricity.

MAGNETIC POLES

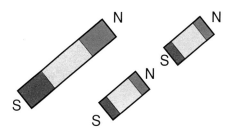

Gilbert's work continued that of Peter Peregrinus, who wrote about magnets in 1269. Peregrinus noticed that magnetic needles pointed towards the Earth's poles, so the ends of a magnet were named north and south poles. He also saw that just as unlike poles attracted, like poles repelled, and that if you cut a magnet in half, you had two magnets instead of one.

ELECTRIC MACHINES

Otto von Guericke (1602-86) was a teacher who lived in the town of Magdeburg in Germany. He knew William Gilbert's work, and tried to test Gilbert's idea that magnetism might be the force of attraction between stars and planets. In 1665, he made a sphere to represent the Earth. At that time people thought the Earth was made largely of sulphur (since this was a material found in volcanoes) so this was the material from which von Guericke made his sphere. The ball was fixed to an axle so that it could be rotated using a hand crank. Von Guericke found that if he let his hand brush the ball lightly as it spun round, it produced fairly sizable sparks. Soon others were finding ways to produce this new phenomenon, which was called 'statical electricity', a term since shortened to static.

▼ British physicist Stephen Gray discovered in 1729 that electricity could travel through certain materials. He is shown here sending an electric charge along a line of thread.

In the 1730s, Charles du Fay (1698-1739), superintendent of gardens for the King of France and a keen chemist, experimented with transferring electric charges to different objects by touching them with a static-charged glass rod. He discovered that sometimes charged objects were attracted to each other, sometimes they were repelled. Du Fay thought that there must be two different electrical 'fluids' – he called them vitreous and resinous electricity – each type attracting the other but repelling itself. He also showed that the human body could be electrified as long as it was insulated, or kept apart, from the ground.

▲ Pieter van Musschenbroek, with a Leyden jar apparatus. He rubbed the sphere to build up a static charge. The jar was charged at the end of the bar.

▶ A more up to date way of storing static charges was developed by American scientist Robert Van de Graaff in the 1930s. Vast charges of electricity can be built up in the metal sphere of the Van de Graaff generator.

INSULATORS AND CONDUCTORS

Conductors are materials that pass electricity easily. Examples include metals such as copper and iron. Insulators are materials that do not pass electricity easily. Examples include glass, wax and rubber.

In 1745 Ewald von Kleist, a priest from Pomerania (now part of Germany) experimented with a device for storing electricity. It was a glass jar part-filled with water and sealed with a cork that had a nail running through the middle, dipping into the water. By touching the nail to a static electricity generator, von Kleist found that he could store the charge in the jar. Touching the nail to another object afterwards produced a strong spark as the jar discharged the electricity it had stored. Pieter van Musschenbroek of Leyden University, Holland, followed up von Kleist's experiments with great success. Soon many scientists were experimenting with the jars, and before long, no self-respecting laboratory was without one, now called the Leyden jar, after the university. The American statesman Benjamin Franklin (1706-90) was one of these investigators. He realized that the spark from a Leyden jar might be connected with some very much larger sparks.

DOWN TO EARTH

When Benjamin Franklin looked at a discharging Leyden jar, it reminded him of a miniature lightning flash. Could the two actually be the same thing, but on vastly different scales? He set down his reasons for thinking that lightning was a form of electricity, among them that it gave out light, it moved swiftly and was conducted by metals. He concluded: 'they agree in all the particulars in which we can compare them... Let the experiment begin'.

In June 1752, Franklin went out with his son into a field near Philadelphia, USA, to fly a kite in a thunderstorm. The kite had a pointed wire attached to it, to which Franklin tied a silk thread that would be charged by any electricity that was overhead. On the end of the silk thread was a key. As the lightning flashed above him Franklin put his hand near the key, drawing sparks from it. He also found that he could charge up a Leyden jar with the key, proving that lightning was a form of electricity. However, Franklin was a very lucky man. When Professor Richmann of St Petersburg, Russia, tried the same experiment his assistant saw, 'a glow of blue fire as large as his fist' flash to the professor's head from his kite, killing him instantly.

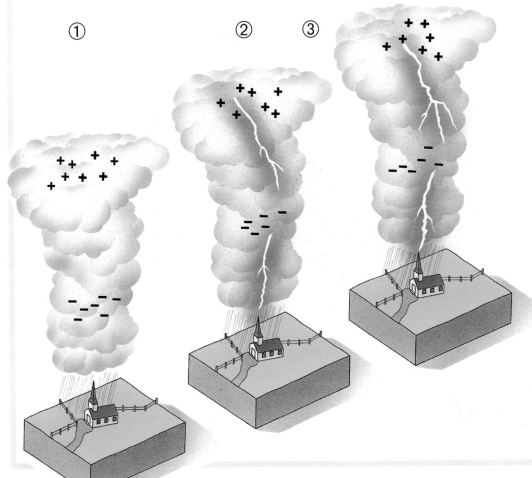

◀▶ The lightning bolt is one of nature's most dramatic spectacles. It occurs when storm clouds gather, and the water and ice particles inside them bump together. As a result, static electricity builds up. Positive charges collect in the highest clouds, and negative charges in the lowest (1). When the build-up of charges is too great, a flash of lightning is released (2) – either from the bottom to the top of the cloud or from the bottom to the ground. Here the church's lightning conductor is struck (3), saving the building itself from damage.

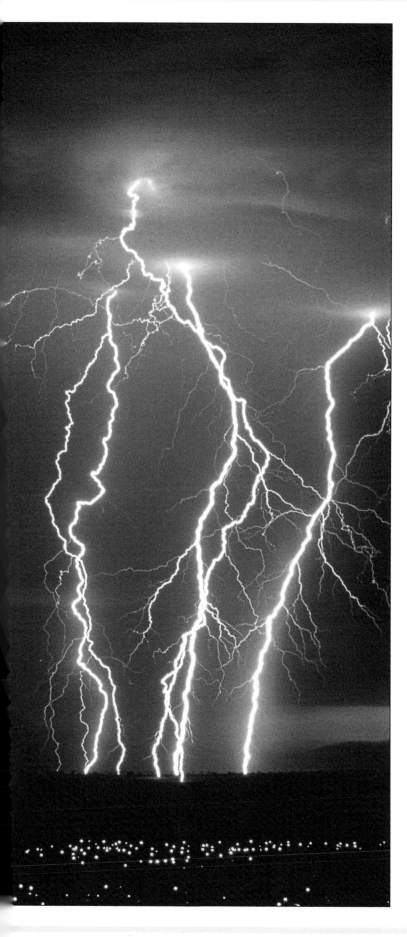

Franklin thought of a practical benefit from his experiments, suggesting that metal rods be attached to the sides of tall buildings with wires leading into the ground. Such a conductor would carry an electrical discharge safely to earth, protecting the building from damage.

Franklin proposed that rather than there being two kinds of electricity, as du Fay had suggested, there was just one. An object with excess electric fluid was 'positive', attracting objects with little electric fluid, which were 'negative'. Electricity would move from the positive to the negative. Later research showed that Franklin, although wrong in thinking electricity to be a fluid, was right in principle.

In 1785 Charles Coulomb (1736-1806), a French physicist, showed that the force of attraction between electrically-charged objects changed according to distance, reducing as the objects were moved apart.

✷ CONDUCTING LIGHTNING

Franklin's experiments, dangerous as they were, provided much useful information as to the nature of lightning and natural electricity. Franklin's lightning conductor (a conductor is anything that makes an easy path for electricity to follow) has become a standard fitting on all high buildings. Lightning rods are usually made of copper, which is an excellent conductor, providing a harmless path for the lightning to follow to earth.
★ *Remember that Franklin's experiment was dangerous. Never be tempted to repeat it yourself.*

GALVANI AND VOLTA

Around 1780, Italian scientist Luigi Galvani (1737-1798) was dissecting a frog on a metal sheet. When he touched one of the dead frog's nerves with his steel scalpel, a leg twitched. Galvani experimented further and found that the frog's legs would twitch whenever they were brought into contact with two different metals, such as brass and iron. He believed that the frog's muscles were producing 'animal electricity', as he called it.

Galvani sent a report of his findings to a friend, another Italian called Alessandro Volta (1745-1827). Volta repeated the experiments and decided that Galvani's explanation was wrong, reasoning that the electricity was a result of the brass and iron being in contact with each other, not by the animal tissue. Galvani was furious at being contradicted, and a bitter row broke out between the two men.

NAMES THAT LIVE TODAY

The principles of Volta's pile allowed the development of the batteries we use today (see page 30). Volta, shown above, likened the steady flow of electricity from his pile to that of a river, calling it a current, and the unit of electrical force that moves this current is named the volt. Galvani's name also lives on today. The electric current set up by two metals in contact with each other is known as galvanic electricity. Iron, a metal that corrodes quickly in damp conditions, can be protected from rusting by coating it in zinc. This process is called galvanizing.

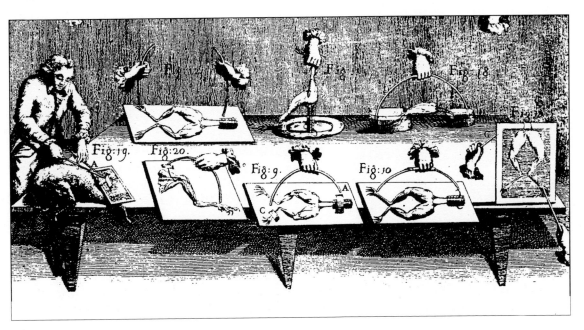

◀ Galvani experimented for many years in his laboratory. This picture is taken from a print made at the time.

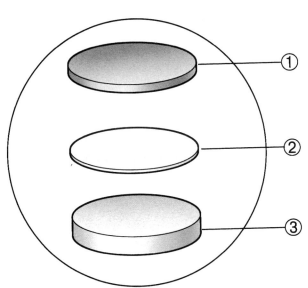
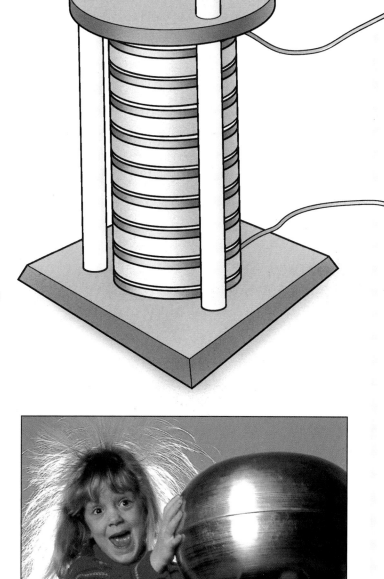

▲▶ Sizes of voltaic piles varied, according to the output required, the biggest consisting of several hundred cells connected in a circuit. The principle is the same, whatever the size. By a chemical reaction, the copper takes up electrons from the salt solution, the zinc loses electrons to the solution. The resulting flow of electrons can be harnessed as an electric current.

1 Copper disc
2 Pasteboard disc soaked in salt water
3 Zinc disc

In 1800 Volta built a device – the 'voltaic pile' – that proved electricity could be produced by a reaction between two different metals. The basic unit of his pile was a cell, made up of a sandwich of copper, pasteboard and zinc discs. The pasteboard was soaked in salty water. Volta built a pile of 30-40 three-disc cells, to form the first battery. When connected up, the voltaic pile produced a steady flow of electricity, a breakthrough discovery, since the Leyden jar released its charge in a single surge.

The invention brought Volta fame and honour. The French emperor Napoleon Bonaparte summoned Volta for a demonstration, and fellow scientists honoured the achievement. The force that moves an electric current, the volt, was named after him.

▲ In the early days of electrical research, people were fascinated by the new force, and many paid to get an electric shock. A popular demonstration involved suspending a boy with silk cords, which acted as insulators, then charging him with static electricity so that his hair stood on end. Here, a girl of today repeats a similar experiment.

LIVING ELECTRICITY

Although Galvani's idea of 'animal electricity' was rejected by Volta, it wasn't entirely wrong either. When placed between two different metals, the chemicals in the body helped create a very weak electrical impulse, acting like the salt solution in the pasteboard discs of the voltaic pile. And when Galvani attempted to charge living tissue from a Leyden jar, the liquids acted as a conductor.

Your body is a complex electrical system. The information received through your five senses – those of sight, sound, touch, taste and smell – is passed along nerve fibres as tiny electrical signals to the brain, which is also an electrical activity centre. These electrical signals are measured in tiny amounts. They travel as messengers and controllers throughout the body, especially in the brain, muscles and sense organs. In the nineteenth century, many curious 'medical' machines were developed, often with no real effect other than to cause mild electric shocks and other odd sensations to patients.

However, research did lead to such important machines as the electrocardiograph, or ECG machine. Developed in 1900 by Willem Einthoven, the ECG detects the tiny electrical signals generated by the heart muscle. The rhythm can be shown as a moving 'blip' on a small TV screen. This can then be used to monitor the progress of a patient in hospital with heart trouble. Likewise, brain waves can be detected by a similar machine, the electro-encephalograph, or EEG.

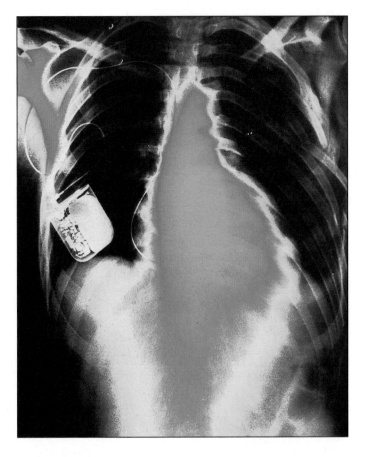

▲ The heart's pumping rhythm is an electrical process controlled by a patch of special tissue. If this becomes faulty, a heart pacemaker (invented in 1960) like the one shown on the left in this X-ray picture may be surgically implanted under the skin. This battery-powered machine, similar in size to an audio cassette, provides the regular signals that the heart needs in order to function properly.

▲ This clock works on the same principle as Volta's pile. Acid in the fruit juice acts as an electrolyte connecting electrodes made of different metals.

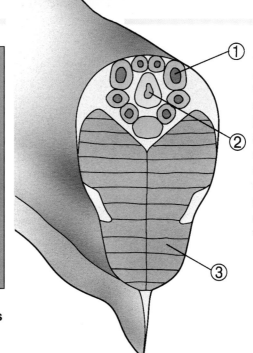

◀ A cross-section through an electric eel's body reveals swim muscles (1), spine (2) and electric organs (3) that can build up a charge of 600 volts or more.

Many creatures in the animal kingdom are well known for using electricity. Electric eels and rays have special muscles which behave like living batteries. These produce pulses of electricity which can discharge through the water to shock victims. Sharks too, use electricity for homing in on their prey. About 1,000 sensors in a shark's skin enable it to detect the minute electrical forces generated by the muscles of nearby fish. The result is that a hungry shark can move in with deadly accuracy, even in utter darkness.

▲ Hundreds of cells in an electric eel's skin act as tiny batteries. They can deliver a massive electrical charge that is enough to stun or kill nearby prey.

ELECTROCHEMISTRY

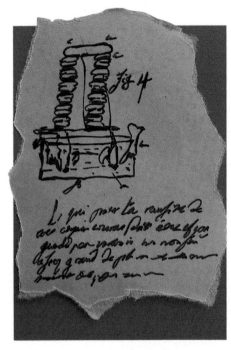

▲ Part of Volta's letter to the Royal Society in London.

Volta wrote to a famous scientific group, the Royal Society of London, to tell them about his invention. One of the members was William Nicholson (1753-1815), an English chemist, who built his own voltaic pile. He placed the ends of the battery wires into water and found that bubbles of gas were given off – the electric current was splitting the water into its two basic elements, hydrogen and oxygen. Volta had shown that a chemical reaction could cause an electric current, and now Nicholson showed that an electric current could cause a chemical reaction.

Sir Humphry Davy (1778-1829), another English chemist, carried out further experiments. He built a battery with over 250 cells, the most powerful yet assembled. In October 1807, Davy applied the current from his new battery to molten potash. The result was astonishing – the electric current caused a chemical reaction in the potash, and drops of a silver-white metal appeared, a pure element that Davy called potassium. Within a week, using a still larger current, he extracted pure sodium from soda and the following year went on to discover four more new metals.

◄ Humphry Davy used a huge battery of cells for his electrolysis experiments.

▲ Electrolysis is used for many industrial processes, among the most popular being chromium-plating parts of automobiles, such as the bumpers and window surrounds.

Davy gave regular lectures at the Royal Institution, founded in 1799 in London. The lectures were generally well illustrated by experiments, a tradition carried on today in the shape of Christmas lectures for young people. Several of Davy's lectures were attended by Michael Faraday (1791-1867), who applied for a job as Davy's assistant. Faraday, perhaps more than any other man, was to carry out researches that would ultimately electrify the world, as we shall see later in this book. Continuing Davy's work in electrochemistry, Faraday gave the name electrolysis to the process of separating out elements. The substance to be split he called an electrolyte, and the metal rods between which the current flowed, Faraday called electrodes. The positive electrode was the anode, the negative electrode the cathode, terms still used today.

 A LABORATORY METAL

A substance such as potassium can be produced only in the laboratory. In nature it is always found combined with other substances as a compound. When separated out by the process of electrolysis, potassium appears as a silvery metal, soft enough to cut with a knife.

Davy and Faraday's work led to the industrial process of electroplating. This process covers a conducting object with a thin coat of pure metal. Silver cutlery is an example: silver spoons are expensive; spoons plated with silver look nice and shiny, but cost a small fraction of solid silver.

ELECTRICITY AND MAGNETISM

By the nineteenth century, much progress had been made in the understanding of electricity. For some time scientists had suspected that there was a link between electricity and magnetism, but what could it be? In 1820 Danish scientist Hans Christian Oersted (1777-1851), professor of physics at Denmark's Copenhagen University, found that the needle on a compass moved when it was brought near a wire carrying an electric current. The discovery made him famous and he was invited to visit a number of science academies across Europe. During his travels he met fellow scientists Davy and Faraday.

▲ Hans Christian Oersted.

▶ Oersted's equipment consisted of a battery (1), a current-carrying wire (2), and a compass (3). When the compass was moved near the wire, the needle swung away from its north-south direction.

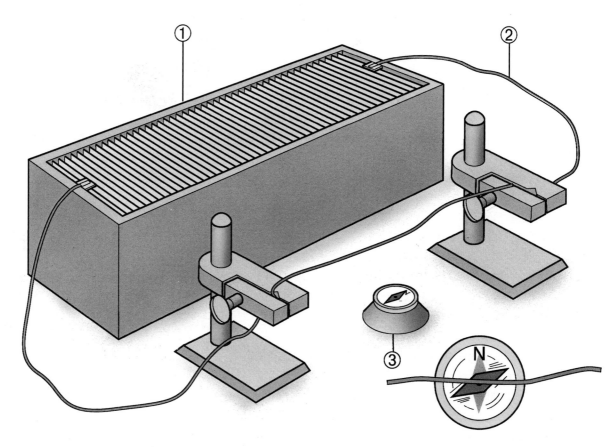

▲ Close-up of magnet's needle.

Shortly after Oersted's announcement, Frenchman André-Marie Ampère (1775-1836) discovered that the movement of the compass needle followed a definite direction. Imagine taking hold of a wire through which current is passing with your right hand, with your thumb pointing in the direction of the current. The north pole of a magnet will always be deflected in the direction of your curling fingers. Ampère also showed that a wire coiled like a spring behaved like a bar magnet when a current was passed through it. He called this a solenoid.

In 1823, following Ampère's descriptions, English physicist William Sturgeon (1783-1850) invented the first electromagnet. His horseshoe-shaped iron bar, surrounded by an insulated wire coil, could lift no less than twenty times its own weight so long as a current was passing through the wire.

▲ This electromagnet is used in a scrap yard for lifting heavy metal parts. It also helps separate the bits containing iron from the rest.

 AMPS

Ampère founded the science of electric currents, which he called electrodynamics. The study of stationary electric charges, which von Guericke and Franklin had worked on, he called electrostatics. The unit that measures the quantity of an electric current passing a point in a given time is called the ampere, or amp.

TELEGRAPH AND RELAY

In America, physicist Joseph Henry (1797-1878) had heard of Sturgeon's work with electromagnets. After experimenting, Henry realized that the more coils of wire there were around the iron core, the more powerful the electromagnet would be, and by 1831 he had made an electromagnet that could lift over 300 kilograms.

Of course, electromagnets did not have to be so powerful, and Henry put one to a much more delicate use. He made a switch that consisted of a 1.6 km-long wire, with a tiny electromagnet at one end and a battery at the other. When the current was switched on the electromagnet attracted a small iron bar, which sounded a bell. When the current was switched off a spring pulled the iron bar back to its original position. With this device, Henry had invented the first practical system of sending simple signals electrically, the telegraph. The bell's on-off sound could be used as a signal code for sending messages.

▲ Joseph Henry worked on electromagnets, electric motors and the electric telegraph.

◄ One of the first telegraph lines ran between Windsor and London.

 RESISTANCE

All conductors resist the flow of electricity to some degree. The difference depends on the material used, its width and length, and the amount of current flow. For a given amount of current, a short, wide conductor has less resistance than a long, thin one. George Ohm summed up the effects in his Ohm's Law, and the unit of resistance is named after him, as the ohm.

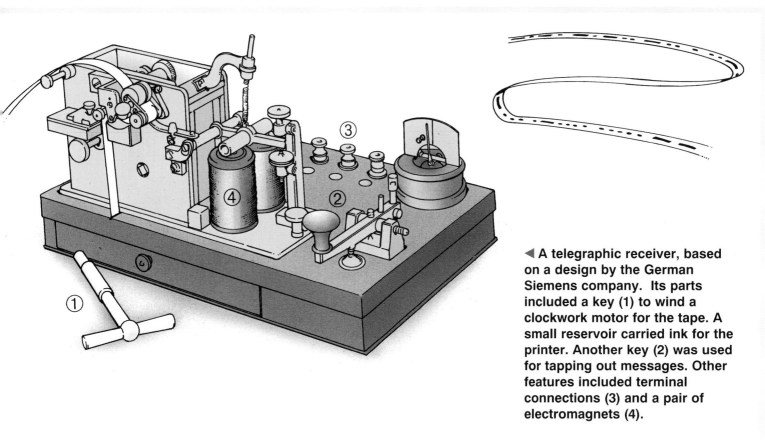

 A telegraphic receiver, based on a design by the German Siemens company. Its parts included a key (1) to wind a clockwork motor for the tape. A small reservoir carried ink for the printer. Another key (2) was used for tapping out messages. Other features included terminal connections (3) and a pair of electromagnets (4).

Henry's telegraph was limited to fairly short distances. In 1826, German physicist Georg Ohm (1787-1854) had shown that different materials opposed, or resisted, the flow of an electric current to varying degrees. Materials that conducted electricity had a low resistance, but this resistance increased as the length of the conductor increased. In order to keep the same amount of current flowing, the voltage had to be increased. The resourceful Henry found a way to get around this difficulty with his electrical relay of 1835. A current just strong enough to switch on an electromagnet could be sent down a wire. The electromagnet closed a switch that turned on a second battery. The current from this battery then travelled along a further length of wire to another relay where the operation was repeated. As many relays as were necessary could be set up to cover great distances, and because each stage length was short, the voltage needed remained low.

Henry did not patent the relay as he believed that scientific discoveries should be freely available for everyone, and when Henry met Samuel Morse he was happy to explain how the relay worked. Morse obtained a patent for the electric telegraph in 1840 and the first telegraph line was opened in 1844, between Baltimore and Washington, USA.

SENDING MESSAGES

The Morse code became the most widely used message system ever, though Morse never admitted how much he learned from Joseph Henry. As the railways were built across the USA, so telegraph wires were erected beside the tracks. This way, apart from other messages, timetable changes or train delays could be flashed between telegraph stations along the route. Later, when radio was invented, Morse code was again put to use as a simple way of transmitting messages. Perhaps the best known part of the Morse code is the emergency SOS (save our souls ... – – – ...) message, still used today.

THE ELECTRIC MOTOR

Oersted's discovery that electricity and magnetism were related led to another important invention. In 1821, the year after Oersted announced his findings, Michael Faraday observed that the force that acted between a magnet and a wire carrying an electric current moved the magnet in a circle around the wire.

He carried out experiments that arranged magnets and wires in such a way that either the magnets or the wires were free to rotate. When a current was passed through the freely-moving wire it would turn around the fixed magnet, or the freely-moving magnet would turn around the fixed wire. Faraday's spinning wires and magnets were little more than toys, but this was the first time that electrical energy had been converted into mechanical energy. It was the first stage in the development of the electric motor.

It was almost ten years later, in 1830, that Joseph Henry, using electromagnets, made the first electric motor that could do useful work. By 1840 electric motors were being used for machines such as drills and lathes, although the batteries needed to run the motors were expensive.

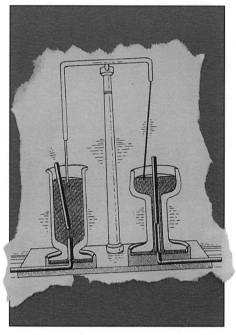

▲ This picture shows one of Faraday's machines. He called it an 'electro-magnetic rotation apparatus'. On the left, a magnet is going around a fixed wire; on the right a wire is spinning round a magnet.

▶ This diagram shows the basic layout of a simple electric motor.
1. Current flows through a coil of wire, called the armature.
2. The magnetic force pushes one side up, the other side down.
3. Mounting the armature on a shaft allows it to spin round.

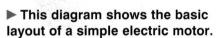
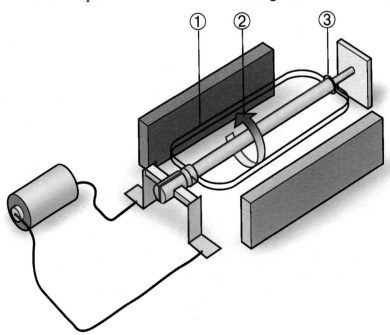

MOTORS BY THE MILLION

Life today is unthinkable without the myriad electric motors, large and small, that perform a multitude of jobs. In a typical three-bedroom house, there are three dozen or more motors whirring away, running everything from microwave ovens and electric clocks to extractor fans and compact disc players. In a luxury car, the motor count runs high too – even doors are sometimes powered.

▲ Early subway systems were a natural for clean electric power.

▲ A lineup of just some of the electric motor-powered gadgets that are used in today's homes.

23

THE GENERATOR

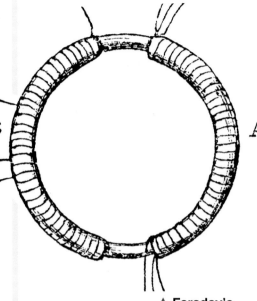

▲ Faraday's induction ring used two coils of wire round a soft iron core. They were kept apart by calico fabric and tied in place with string.

The electric motor produces movement from electricity; the generator produces electricity from movement. The first generator was developed by Michael Faraday in 1831, during the course of a series of experiments.

Faraday wound two separate coils of wire around an iron ring. He linked the first wire to a galvanometer, an instrument for measuring electric current. The second wire was connected to a battery. The wires were carefully insulated so that they had no contact with each other, yet when the current in the second wire was switched on and off the galvanometer registered a current in the first wire. This was caused, or induced, by the variation in the magnetic field produced by current in the second wire.

Faraday thought – correctly – that magnetic 'lines of force' extended out when an electrical current was flowing, collapsing in again when it was switched off. It was the lines of force cutting across it that induced the electric current in his second wire.

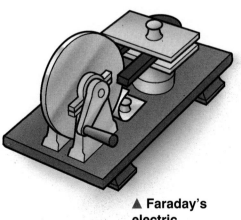

▲ Faraday's electric generator used a copper disc, spinning between the poles of a horseshoe magnet.

The discovery of this phenomenon – electromagnetic induction – was crucial. Faraday went on to demonstrate that an electric current was also produced when a coil was moved near a magnet, or a magnet moved near a coil. In a classic experiment, Faraday placed a copper disc, which could be rotated using a crank, between the poles of a horseshoe magnet. One contact wire trailed against the shaft running through the disc and a second wire trailed against the disc itself. Both wires were linked to a galvanometer. When the disc was turned through the magnet's lines of force the galvanometer registered a flow of current. With this device, Faraday had turned mechanical energy into electricity and perfected the first generator. By the 1860s, electricity generators were in widespread use. These large generators used electromagnets and were driven by steam engines. Improvements in generator design made it possible to produce electricity on a huge scale.

▲ Michael Faraday, shown here lecturing at the Royal Society, was a largely self-taught experimenter, and one of the greatest electrical pioneers. He is probably best remembered for the generator, but his discoveries ranged across the fields of chemistry and magnetism, too.

◀ The generating equipment in a modern power station.

GENERATING POWER

Generators are the foundation of modern life. Without them, power stations (see page 28) would not be able to produce the massive amounts of electricity needed to run homes, offices and factories. The source of energy used to turn such generators varies widely. The range includes natural resources such as coal, oil, gas, wind and hydro power, and nuclear energy.

LIGHTING UP THE WORLD

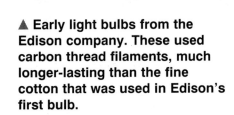

The familiar light bulb we use today was not developed until 1878. Before then, the carbon-arc lamp was the nearest thing to a practical electric light. It was an 1808 invention that used a pair of close-set carbon rods with a powerful electric current passing between them. The spark that jumped across the gap produced a brilliant white light, but the rods burned away in the intense heat and had to be replaced every few hours. Many improvements were made to carbon-arc lights – their bright output was especially useful in lighthouses – but the incandescent system was the type that evolved into the light bulbs we know today.

The first incandescent bulbs (the name means 'glowing with intense heat') were developed independently by Sir Joseph Swan (1828-1914) in Britain and Thomas Alva Edison (1847-1931) in the US. The key to success was to put the filament – the thin wire that glowed – in a near-vacuum. This stopped it burning away too quickly. Swan was first to demonstrate a practical bulb, in December 1878. When the Edison – perhaps the most prolific inventor there has ever been – announced earlier the same year that he would try to find a way, the value of shares in US companies supplying gas for lighting fell at once, because people had such confidence in his abilities.

▲ Early light bulbs from the Edison company. These used carbon thread filaments, much longer-lasting than the fine cotton that was used in Edison's first bulb.

▼ The secrets of a light bulb revealed. The bright glow is caused by the high resistance of the very long and very thin wire. The bulb is not very efficient however, as much energy is wasted as heat.

1 Coiled-coil filament packs more wire in the bulb for more light output. Tungsten is the most commonly-used material today, as it has a very high melting point.
2 Glass bulb.
3 Fine wire frame supports the filament.
4 Bulb filled with argon gas, which prevents the filament burning away.

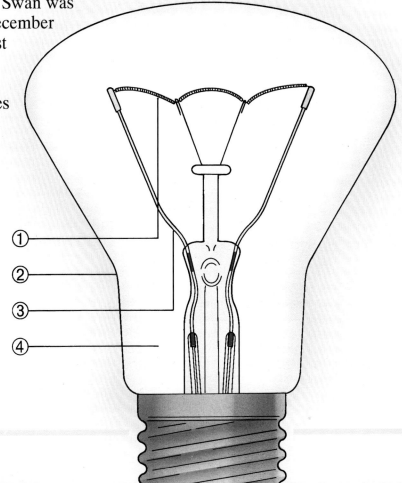

▲ Millions of light bulbs are needed to illuminate the skyline of this city.

After more than a year's work and thousands of experiments, Edison produced his first electric light bulb in October 1879. Two years later, he and Swan teamed up and went into business as the Edison and Swan United Electric Light Company. By the 1890s whole cities were lit by electricity and four million bulbs a year were coming off the production lines. At the same time, vast improvements in design were made – an early bulb lasted 45 hours or so; by 1900, 500 hours was not uncommon. Today, the average light bulb can last thousands of hours.

Gas-discharge lamps were first developed in 1909 by Frenchman Georges Claude. He used a tube containing neon gas, which glowed when an electric current was discharged through it. The bright light uses little current and remains cool, wasting little energy as heat, unlike incandescent bulbs. Such gas-discharge lamps are widely used for advertising, as the tubes can be bent into all sorts of shapes. Different gases are mixed with neon to produce different colours. Street lighting often uses sodium vapour, which gives a yellow, fog-piercing light, or mercury vapour, which gives a hard blue-white light.

The latest energy-saving bulbs are coated inside with a material that glows. They are more expensive to buy than traditional incandescent bulbs, but last much longer and are cheaper to run.

 **FAREWELL TO DARKNESS**

Electric lighting has been a real world changer. Apart from simple things, like being able to read and write comfortably whatever the time of day or night, electric lighting has helped to make the world a safer place. Imagine cars and trucks without powerful headlamps or city streets without powerful overhead lighting systems.

ELECTRICITY FOR ALL

The electric light, and the other electric appliances that followed, needed a reliable source of electricity if they were to be of much use. That meant efficient generators and a means of supplying the electricity to the consumers.

Edison's first power station was built in Pearl Street, New York in 1882. It consisted of three generators, supplying electricity to 225 houses. As demand for electricity increased, for cookers and heaters in homes and electric motors in workshops and factories, so the efficiency of power stations and distribution networks had to be improved. Ohm's law set a limit on the ability of cables to carry electric current – it was impossible to extend the cables over more than a few kilometres.

The solution to the problem was two-fold: alternating current and the transformer. Up to this point power had been supplied by direct current, in which the charge flows in just one direction. In 1887, Nikola Tesla (1856-1943), a former employee of Edison, patented a generating system that used alternating current in which the current flow reverses direction around 50 times a second. It makes no difference to an electrical device which way the current is flowing; it will still work as long as there is movement of electrons.

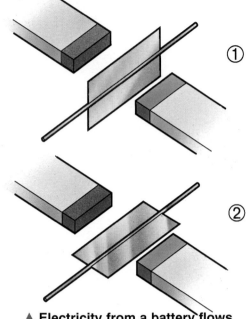

▲ Electricity from a battery flows in one direction and is called direct current. Electricity from a power station is called alternating current – electrons move to and fro in a wire instead of in one direction
1 Power station generators have coils like that of an electric motor. As a coil is turned between two magnets, current is made.
2 The amount of current varies as the coil rotates, from two peaks of current to none at all, in a rapid on-off flow of 50 two-way cycles a second.

▶ An early power station. Most were fuelled with coal, hauled by railway train or barge. Power stations were, and still are, built near a convenient fuel supply.

▲ One of the less glamorous aspects of power generation is pollution. Unless proper cleaning equipment is installed, a coal-fired station creates acid rain, affecting plant life downwind. Even waste steam from the cooling towers can cause local weather changes. With nuclear power stations, the safe disposal of dangerous waste will be a concern for centuries to come.

An alternating current produces a magnetic field that switches on and off rapidly. This is where the transformer comes in, as it can change the voltage of an alternating current. The current flows into a coil wound around the iron core of the transformer. By the process of electromagnetic induction, current is induced in a second coil (Michael Faraday's double-coil device of 1831 was actually the first transformer). The voltage of the induced current depends on the ratio of the turns of the coils. If the second coil has five times as many turns as the first, the voltage goes up five times. Similarly, if there are fewer turns in the second coil the voltage is reduced.

The beauty of the system is that using by alternating current, voltage can be stepped up at power stations for transmission along cables, then stepped down again for use in homes. Electricity could now be distributed over long distances.

PORTABLE POWER

The mains supply is convenient for fixed electrical equipment such as cookers, refrigerators and room lighting. When you are on the move, however, it is essential to have a portable power source.

The most familiar way of powering gadgets such as torches and portable stereos is by using the familiar tube-shaped dry cell. This is based on a design invented in 1865 by a French engineer called Georges Leclanché. The dry cell consists of a zinc case and a central carbon rod, which are the electrodes, separated by an ammonium chloride paste, acting as the electrolyte. The paste is used instead of liquid, to make the cell spillproof. If you have to put more than one dry cell into your torch then you have a 'battery' of cells, linked together in the same way as the cells in Volta's pile of 1799. People rarely talk about dry cells though – to most of us, a dry cell is a battery, whether we use one or several.

▲ Camcorders use rechargeable batteries, typically nickel-cadmium types.

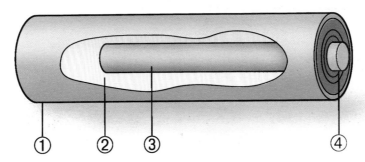

The first rechargeable battery was invented by a French physicist, Gaston Planté, in 1859. It was very similar in design to the liquid batteries that are used today in cars. A series of lead and lead oxide plates are immersed in an electrolyte solution of sulphuric acid and water. A chemical reaction between the acid and the lead plates, in which the plates are converted to lead sulphate, causes current to flow, supplying electricity for starting the motor. Once a car is on the move, a type of generator called an alternator recharges the battery by applying current to the plates. The plates are converted back to lead and lead oxide ready for reuse. Such batteries are good for about three years before replacement.

▲ This type of dry cell is typical of the sort used in small appliances. Various materials can be used. Mercury is common on small hearing-aid type cells.
1. Zinc case
2. Chemical paste
3. Carbon rod
4. Brass end cap

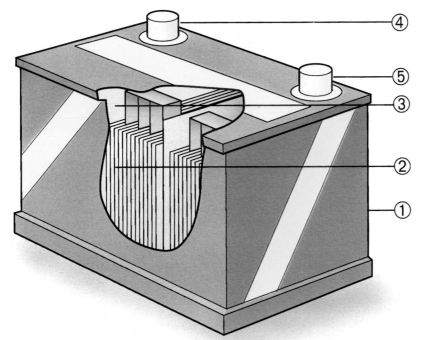

◀ The wet-cell battery is a type that is commonly used in cars and trucks, for starting up and running the electrics while the engine is off.
1. Sealed case
2. Plates
3. Electrolyte
4. Positive terminal
5. Negative terminal

Rechargeable nickel-cadmium dry cells were used industrially as early as 1909. Improvements in recent years have made them suitable for home use as well, and they are used in anything that needs truly portable power, from camcorders to cordless screwdrivers.

With pollution from petrol-fuelled cars a major issue, companies in Europe, Japan and the USA are developing electric cars powered by advanced batteries. They're not perfect yet though – a typical performance is that of the American GM Impact, which can hit a top speed of 160 km/h but is limited to a 240-km range before needing a recharge.

Another source of portable power is the fuel cell, invented in 1959. It converts chemical energy directly into electrical energy. In the most common type, hydrogen and oxygen gases are used as fuels, with hot potassium hydroxide as the electrolyte. As long as there is a continuous flow of the two fuels, the cell provides electricity, with pure water as a by-product, making it ideal for use in manned spacecraft.

▲ The Space Shuttle Orbiter uses three fuel cells for electricity while in space. They supply about the same amount of power that is used by about twenty-four houses back on Earth. The waste water produced by the fuel cells is used by the astronauts for drinking, cooking and washing.

ELECTROMAGNETIC WAVES

▲ James Clerk Maxwell wrote his first scientific paper at fifteen, when he was still at school. His theories of electromagnetism form the basis of much of our present knowledge.

James Clerk Maxwell (1831-1879) was a Scottish mathematician and physicist who linked the findings of Oersted, Faraday and other scientists into a group of equations that summarized all that was known about electricity and magnetism. Maxwell showed that they did not exist apart but were bound together. A magnetic field moving back and forth, or oscillating, gives rise to an oscillating electric field, which in turn produces a magnetic field, and so on. The fields keep in step with each other as an electromagnetic 'wave', moving outwards from a source at the speed of light, about 300,000 kilometres per second.

It was not until the 1880s, after Maxwell's death, that German physicist Heinrich Hertz (1857-1894) proved that Maxwell's wave theory was right. In one of Hertz's early experiments he used a pair of high-voltage coils, with a gap between them. When the power was switched on, a spark leapt across the gap, the rapidly oscillating electric current in it producing an electromagnetic wave. As a receiver, Hertz used a pair of rods with a gap between them. When the waves were picked up, a spark crackled between the rods. For some time these waves were called Hertzian waves, but after they were put to practical use by the Italian Guglielmo Marconi for transmitting messages, they became known by the name we use today – radio waves.

▶ The electromagnetic spectrum is the name given to the range of different types of radiation. The part of the spectrum that our eyes can detect, we call visible light. We can feel infra-red as heat, and too much ultraviolet causes sunburn and skin cancers. Microwaves are used in domestic ovens, radar equipment and communications. X-rays are used in medicine to take pictures of the inside of the body, showing bones and body tissues beneath the skin.

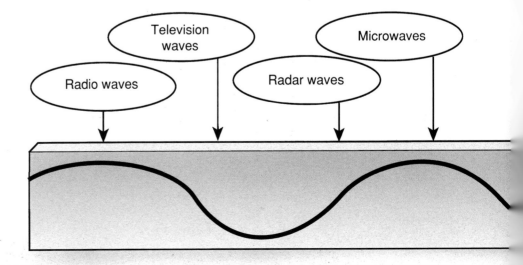

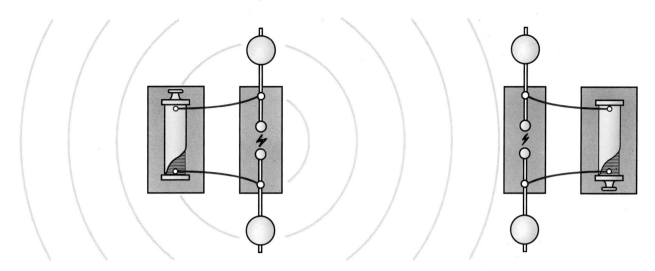

▲ The experimental apparatus of Heinrich Hertz generated a powerful electric spark. As this jumped the gap between two rods, it created a burst of radio waves that moved outwards, like ripples in water when you throw a stone. The receiving apparatus picked up the radio signals, which in turn created a smaller and weaker spark.

 PIONEER OF THE RADIO AGE

From Hertz's pioneering work, the radio has gone on to conquer the world of communications. Guglielmo Marconi developed Hertz's basic ideas, first in Bologna, northern Italy, then later in Britain. Marconi had wireless telegraphy equipment working reliably by the 1890s. This was a half-way stage to radio proper – messages were sent using Morse code, but without needing the wires of a conventional telegraph setup. By 1901 Marconi was sending messages across the Atlantic Ocean and before long, regular broadcasts of news and entertainment were filling the air waves. By the 1930s, a radio set in the home was common, and families sat round the radio to listen to broadcasts in much the same way they would later watch television programmes.

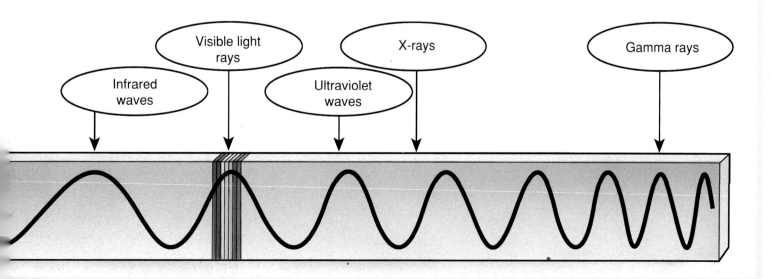

THE CATHODE-RAY TUBE

In the 1850s, German scientist Julius Plücker noticed a mysterious green glow surrounding the cathode in a vacuum tube with which he was experimenting. Plücker was not the first to have spotted this – in 1838, Michael Faraday had noticed the effect, but the vacuum tubes at the time were not good enough to investigate further. As vacuum tubes improved, researchers found that the glow was in fact a stream of rays. In experiments with these cathode rays, the British scientist William Crookes (1832-1919) put a small metal cross inside a vacuum tube. When the power was switched on, the end of the tube glowed green where the cathode rays hit it, while in the middle the metal cross cast a shadow, proving that the rays moved in straight lines. Experiments showed that they could be bent, or deflected, by magnets placed nearby.

▲ William Crookes perfected his cathode-ray tube in 1880.

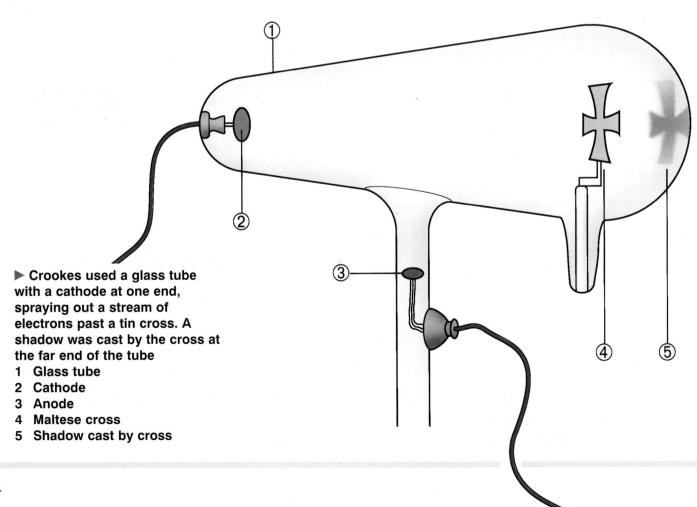

▶ Crookes used a glass tube with a cathode at one end, spraying out a stream of electrons past a tin cross. A shadow was cast by the cross at the far end of the tube
1. Glass tube
2. Cathode
3. Anode
4. Maltese cross
5. Shadow cast by cross

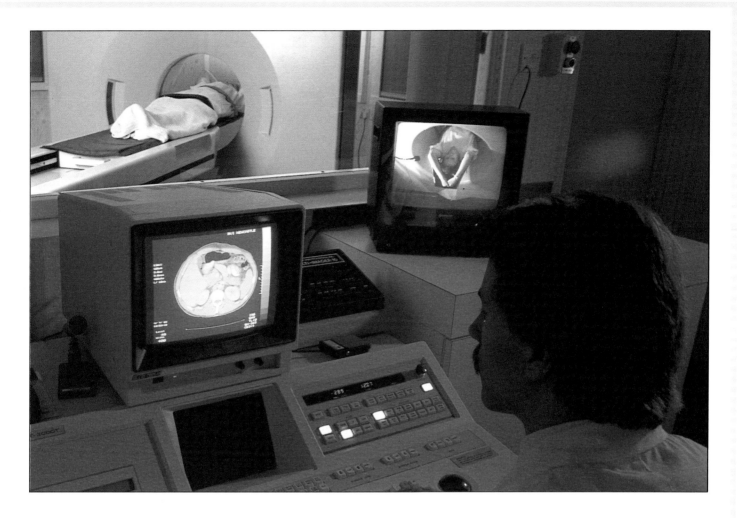

▲ Modern cathode-ray tubes, here used as monitors for a hospital body-scanner system. High quality equipment provides an excellent picture, though eye strain is a common complaint among workers who use such screens for long periods.

Another British researcher, Joseph John Thomson (1856-1940) knew that cathode rays could be bent by a magnetic field and he carried out work that proved the rays were negatively charged. In 1881, he discovered that the rays were made of particles that were far lighter than even the lightest atom, hydrogen. This was a breakthrough – until this time, people had thought the atom was a single unit that could not be split up.

Thomson's research showed that, in fact, the atom had to be made up of smaller components. At first, Thomson called these particles 'corpuscles', but eventually they were called electrons. This was a word that was used some years earlier by the Irish physicist George Stoney, after the original Greek word for amber.

 ELECTRONIC DISPLAYS

From the work of the researchers on cathode rays has come our present knowledge of the atom and its structure. The cathode-ray tube itself, now shortened usually to CRT, has been developed into the familiar picture tube used in TVs, radar screens, computer monitors and all the other electronic displays that are used every day across the world.

THE ELECTRONIC AGE

Electronics, the study of devices that rely on the controlled flow of electrons, dates from 1904, with British electrical engineer Sir John Fleming's new invention, the diode valve. It turned on when current flowed in one direction, off when the flow went the other way. Valves, with their one-way current flow, made it possible to build radios that could receive weak broadcast signals. Two years later, the American Lee De Forest developed the idea by making a valve that could increase, or amplify, current as well. De Forest's improved design, the triode valve, meant that a loudspeaker could be used for listening to a radio instead of headphones.

Over the next thirty years valves were put to many other uses, in the first televisions and computers, for example. They were bulky and fragile, however, and people looked for something smaller and tougher that could do the same job. In 1947, a three-man team led by William Shockley at the US Bell Telephone Laboratories invented the transistor.

▲ This early computer dwarfs today's equipment. Despite this, an average modern desktop machine is much more powerful.

▶ Key stages in the development of electronic components included delicate glass valves, which came in many shapes and sizes. A transistor (far right) does the job of a valve, while being much smaller and more reliable. Today's integrated circuits pack thousands of transistors on a tiny piece of silicon chip. Each stage of size reduction has improved speed, power and reliability many times over.

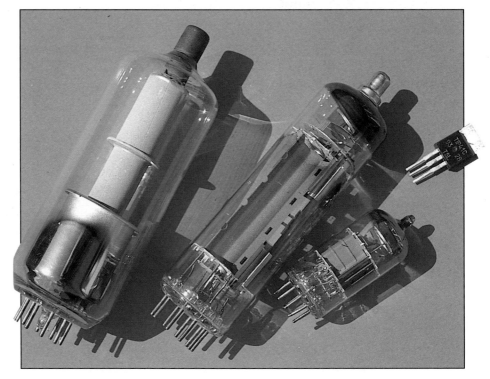

▶ This closeup of an electronic circuit shows how complex such devices can be.

A transistor depends on materials called semiconductors. For as long as electricity had been studied it was known that some solids would conduct electricity and some would not. However, there were elements, such as silicon and germanium, that fell somewhere in between. These are called semiconductors. They can be produced either rich in electrons (n-type) or which attract electrons (p-type). By putting p-type and n-type semiconductors together in particular ways, it is possible to make devices that can switch, amplify or detect electric current.

The transistor had huge advantages over the old-fashioned and bulky valve – it was lighter, smaller, more reliable and a lot cheaper to make. This meant that electronic equipment could be made much smaller, and the first transistor radios used the new technology with great success.

The next leap in the evolution of electronics came when Americans Jack Kilby and Robert Noyce developed the integrated circuit, in 1958. This was a way of having several components on a single piece of semiconductor material. Previously, transistors had been linked individually to other components in an electrical circuit, just as valves had been. By the 1980s it was possible to have a million or more transistors on a single silicon microchip. The result has been an explosion in the use of machines that use them – from desktop and laptop computers, to portable telephones and satellite navigation systems.

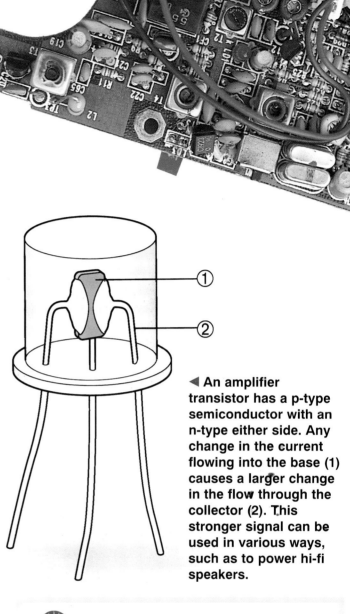

◀ An amplifier transistor has a p-type semiconductor with an n-type either side. Any change in the current flowing into the base (1) causes a larger change in the flow through the collector (2). This stronger signal can be used in various ways, such as to power hi-fi speakers.

ELECTRONIC PARTS

Transistors are not the only semiconductor parts in an electronic circuit. Others include: resistors, which control the amount of current flowing in a circuit; diodes, which allow current to pass in one direction only; and light-emitting diodes (LEDs) which glow when current passes through. LEDs are used for indicator lights and in some calculators. Capacitors store an electric charge, releasing it when required, such as when a camera uses an electronic flash.

ELECTRICITY FROM LIGHT

When Hertz was conducting radio-wave experiments in 1887 he noticed that the sparks produced in his detector were influenced by light falling on the spark transmitter. He saw that the effect was greatest when light fell on the transmitter's cathode, but did not take his investigations much further.

Other scientists discovered that light was causing the cathode to give off electrons, but they had no explanation for why this should be so. An answer came from the German-born scientist Albert Einstein (1879-1955). In 1905 he proposed that light, and all other forms of electromagnetic radiation, could sometimes act as particles rather than waves. When a particle of light, called a photon, struck an electron in a solid, the electron absorbed the photon's energy. If the photon had enough energy, the electron was ejected from the solid. Einstein was saying that electromagnetic radiation could act as a wave or as a stream of particles.

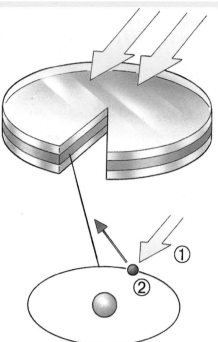

▲ The photoelectric effect is at work in a solar cell or photocell, which turns light into electricity. Energy from photons (1) frees electrons (2) from a material such as silicon and the resulting current flow can be used as a source of power. Solar cells are in wide use today, from powering calculators to supplying electricity aboard satellites.

▲ Down on Earth, uses for solar cells range wider every day. This Honda Dream rolled to victory in the 1993 trans-Australia World Solar Challenge race. The car averaged 84 km/h over the 3,000 km course, hitting speeds of up to 112 km/h. The Dream's solar cell array charged 83 silver-zinc battery cells, which in turn powered an electric motor.

Two other ways of producing electricity are through heat and pressure. Heating one junction of two conductors results in a current flow. It is known as thermoelectricity, and was originally studied by German scientist Thomas Seebeck (1770-1831) in the 1820s. Piezoelectricity is that produced by pressure. Pressing some crystals, including quartz, results in a current flow. Devices that use the effect include the hand sparkers often used to light the cooking rings on gas ovens, and the quartz watch.

 A SOLAR FUTURE

Solar cells don't catch the headlines, but they stand a good chance of being a leading power source in the future. Constantly improving technology means that weight and size are shrinking, while power output goes up. It's in the small ways that solar cells are creeping into our lives – whether it is a laptop computer that is freed from dependence on the mains plug, or the solar-powered air-conditioner that keeps a car interior cool while the petrol engine is switched off.

◀ **Solar cells cover the surface of this satellite. As long as the craft remains in sunlight, electricity can be generated to run all the systems.**

NEW FRONTIERS

At room temperature, even the best conductor shows some resistance to an electric current passing through it. However, back in 1910, the Dutch physicist Heike Kamerlingh Onnes (1853-1926) was researching the behaviour of matter at very low temperatures. He discovered that the resistance of materials to electrons passing through them dropped off to zero at temperatures close to -270°C. He called this astonishing property superconductivity.

As we saw earlier, one of the problems in transporting electricity from generator to user is the high voltages needed to overcome resistance in the wires and cables. If superconducting wires could be used, the cost of supplying electricity would be greatly reduced. However, in most materials superconductivity only appears at ultra-low temperatures and the expense of the cooling systems required would be far greater than any savings made.

RESEARCHING THE FUTURE

In Japan, scientists are working on a project involving 'quantum wires'. These would be so tiny that a computer rivalling today's wardrobe-sized Cray supercomputer could be the size of a book.

Elsewhere scientists are working on special types of plastic that conduct electricity. These would revolutionize many areas – a portable radio for example, could have its battery built into the casing.

◄ A magnet floats in mid-air above a ceramic super-conductor. The vapour is from liquid nitrogen, used to keep the material cool.

◀ The MLU-002 maglev is used for developing high speed transport links in Japan. In Europe, the German Transrapid magnetic hovertrain is a cutting-edge design that is ready to catapult commuter speeds to 400 km/h and beyond. A 283-km route between Berlin and Hamburg should be in operation by the year 2002. The journey will take just fifty-three minutes. The train relies on research into superconductivity for its advanced design.

Ever since Onnes' discovery, scientists have searched for materials that are superconducting at higher temperatures. In 1988 superconductors that worked at around -150°C were found. This is still very cold, but components can be cooled to this temperature with liquid nitrogen, which is not too expensive to make. Scientists are now trying to apply the technology to computers – using superconducting circuits, such computers would be hundreds of times faster than anything that is available today. And a French team claims to have discovered superconductors that will work much nearer to 0°C, which would really revolutionize things.

In the course of 400 years of research scientists have established that electricity involves the movement of electrons. If an object has an accumulation of electrons, it will have an electric charge. But what moves the electron along? The physicist Richard Feynman (1918-1988) and others have proposed that electrons exchange even smaller, massless, chargeless particles to propel themselves, but really the answer is not yet known. Could you be the one to find it?

TIMELINE OF ADVANCE

Here are some of the people whose science discoveries, inventions and improvements have brought about the world of today.

Thales of Miletus Greek philosopher (624 BC-546 BC). Considered to be the father of modern science. The first person we know of who studied magnetism, and the first to ask, 'what is the universe made of?'

William Gilbert English doctor and scientist (1544-1603). Carried out many experiments on electrical and magnetic forces. His book *De Magnete* set the scene for the scientific study of electricity and magnetism.

Otto von Guericke German physicist (1602-86). Famous for his experiments involving vacuums. Devised the first machine for generating an electric charge.

Charles du Fay French physicist (1698-1739). Studied the ways in which objects with an electric charge attract and repel each other.

Benjamin Franklin American statesman and scientist (1706-90). One of the authors of the American Declaration of Independence, 1776. Proved the electrical nature of lightning by flying a kite in a thunderstorm.

Charles Coulomb French physicist (1736-1806). Inventor of instruments for measuring electric charges and the forces between magnets. The coulomb, the unit of electric charge, is named after him.

Luigi Galvani Italian anatomist (1737-98). Accidentally discovered that two metals could react together to give an electric current but did not recognize what he had found. The galvanometer, a device for detecting electric currents, is named after him.

Alessandro Volta Italian physicist (1745-1827). A friend of Galvani who correctly interpreted his findings and used this knowledge to make the first battery, his voltaic pile. The volt, the unit of electrical force, is named after him.

William Nicholson English chemist (1753-1815). Built his own voltaic pile and showed that electricity could be used to break up chemical compounds.

André Marie Ampère French physicist and mathematician (1775-1836). Founder of the science of moving electric currents, called electrodynamics. The unit of electric current, the ampere or amp, is named after him.

Hans Christian Oersted Danish physicist (1777-1851). Discoverer of electromagnetism.

Sir Humphry Davy English chemist (1778-1829). Discovered a number of metals, including sodium and calcium, by using electrolysis to extract them from their compounds.

Sir Humphry Davy

William Sturgeon English physicist (1783-1850). Inventor of the electromagnet.

Dominique Arago French physicist (1786-1853). Followed up Oersted's discovery of electromagnetism and showed that metals other than iron could be magnetized.

Georg Simon Ohm German physicist (1787-1854). Defined the law that showed the relationship between current, voltage and resistance in an electric circuit. The unit of resistance, the ohm, is named after him.

Michael Faraday English chemist and physicist (1791-1867). Despite having no formal education, he became one of the greatest scientists there has ever been. Inventor of the electric motor, the electric generator and the transformer, and discoverer of electromagnetic induction.

Joseph Henry American physicist (1797-1878). A scientist on a par with Faraday who made many of the same discoveries independently. Inventor of the electric relay and the first practical electric motor.

Sir Joseph Swan English physicist and chemist (1828-1914). Inventor of an electric light bulb that rivalled that invented by Thomas Edison. Swan formed a joint electric company with Edison in 1883.

James Clerk Maxwell Scottish mathematician and physicist (1831-79). First person to describe electricity and magnetism mathematically and predict the existence of electromagnetic waves. Also took the first colour photograph in 1861.

Gaston Planté French physicist (1834-89). Inventor of the rechargeable battery.

Georges Leclanché French engineer (1839-82). Inventor of an electric cell that is the basis of the batteries we use today.

Thomas Alva Edison American inventor (1847-1931). A prolific inventor who took out almost 1300 patents. Invented the first practical electric light bulb, the phonograph and showed the first moving pictures on a strip of film.

Sir John Fleming English electrical engineer (1849-1945). Inventor of the valve for controlling the flow of electric current.

Nikola Tesla Croatian-American engineer (1856-1943). Inventor of the alternating current system by which electricity is sent efficiently from power station to consumer.

Sir Joseph John Thomson English physicist (1856-1940). Discoverer of the electron. Electrons are present in all atoms and are the fundamental units of electric charge. When asked what one looked like he replied, 'It's a red-nosed pixie chasing other pixies around'.

Heinrich Hertz German physicist (1857-94). Discoverer of radio waves. The unit of wave frequency or cycle, the hertz (Hz), is named after him. One hertz equals one cycle per second.

Lee De Forest American inventor (1873-1961). Improved Fleming's valve, enabling it to amplify current as well as direct it. Also demonstrated the first moving pictures with sound in 1923.

Heinrich Hertz

Guglielmo Marconi Italian scientist (1874-1937). Developed early radio broadcasting, the first long-distance communications system that did not require wires.

Albert Einstein German-American physicist (1879-1955). More than anyone else, Albert Einstein's ideas have shaped physics in the twentieth century. His discovery of the photoelectric effect and his theories of relativity changed absolutely the way we believe the universe works. By showing that mass and energy were interchangeable he pointed the way towards harnessing the power of the atom.

William Shockley English-American physicist (1910-1989). Led the team of scientists that invented the transistor, opening the way for compact electronic devices.

43

GLOSSARY/1

A ready-reference guide to many of the terms used in this book.

Alternating current An electric current that reverses its direction again and again, very rapidly. Often abbreviated as AC.

Amber A yellowish substance that is the fossil form of the resin that oozes from some trees. Amber acquires an electric charge when rubbed. The word electricity comes from the Greek word for amber, *elektron*.

Ampere The unit of electric current. It measures the rate of flow of an electric charge. Usually shortened to amp.

Battery Two or more electrical cells in series. A battery stores chemical energy which can be converted into electrical energy.

Catalyst A substance that changes the rate of a chemical reaction without being changed itself.

Cathode rays Streams of electrons from the cathode in a vacuum tube.

Cell A unit for producing electricity in which two electrodes are in contact with an electrolyte. Chemical reactions between the electrodes and the electrolyte produce an electric current.

Compound A substance which consists of atoms of more than one element.

Direct current An electric current that always flows in the same direction, such as that which comes from a battery. It is often abbreviated as DC.

Dry cell A cell in which the electrolyte is in paste form. The batteries commonly used in radios, calculators and other portable gadgets are dry cells.

Electric charge A property that gives rise to electrical and magnetic effects; it may be positive or negative. Like charges repel, opposite charges attract.

Electric current The flow of electrons through a conductor. It may also be used to mean the rate at which an electric charge flows past a point (*see ampere*).

Electrode One of the two parts of an electrical circuit that are in contact with the solution during electrolysis. The positive electrode is called the anode, the negative electrode the cathode.

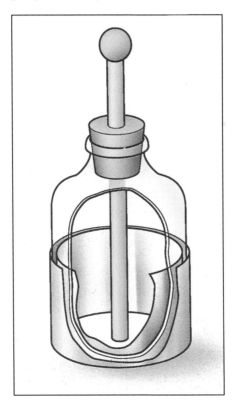

◀ **This cutaway diagram shows one form of Leyden jar. It consists of a glass jar with tin foil coatings, inside and outside, on which an electric charge can be stored. A brass rod contacts the inner foil, to transfer a charge into the jar. The charge can then be stored until the brass rod is connected to a conductor. A rubber stopper prevents the rod touching the neck of the jar.**

Electrodynamics The science of moving electric currents.

Electrolysis A process by which a compound is broken down into its elements by passing an electric current through it.

Electrolyte A substance that increases the ability of a liquid to conduct electricity when it is dissolved in it.

Electromagnet A magnet consisting of a soft iron core around which a coil of insulated wire is wound. When a current is passed through the wire the core becomes magnetized.

Electromagnetic induction Production of an electric current in a conductor by a changing magnetic field.

Electromagnetic wave Radiation consisting of electric and magnetic fields moving together at right angles to one another. Electromagnetic waves travel at the speed of light, about 300,000 km/h. Radio waves, visible light and X-rays are all examples of electromagnetic wave.

Electron A tiny, negatively-charged particle that forms part of every atom. Electrons move in orbits around an atom's central nucleus. Free electrons form current flow.

Electrostatics The study of electric charges that are at rest and the forces that act between them.

◀ An ultra-close look at a light bulb filament. The tight coils enable more filament to be packed into a given length, so allowing a brighter light.

Filament The thin wire inside a light bulb that glows when a current is passed through it. Filaments are usually made of the metal tungsten for its high heat resistance – tungsten can reach 3,400°C before it melts.

Fuel cell A cell in which chemical energy is converted into electrical energy. In the simplest type, hydrogen is combined with oxygen to form water, releasing electrons in the process to produce an electric current. Used aboard the Space Shuttle Orbiter.

Galvanic electricity A steady flow of electric current.

Galvanometer An instrument for detecting and measuring small electric currents.

Generator A device that changes movement energy into electrical energy.

Insulator A substance that is a poor conductor of electricity and stops it from flowing. Typical examples are rubber and most plastics.

Integrated circuit A miniature electronic circuit produced on the surface of a single semiconductor, such as silicon. They are widely used in electronic equipment such as computers and calculators.

Leyden jar Device for storing an electric charge, invented in 1745.

Lines of force Imaginary lines in a magnetic or electric field that allow us to picture the strength and direction of the field.

GLOSSARY/2

Lodestone *See magnetite.*

Magnetic field The field of force associated with the magnetism of a magnetic object. The magnitude and direction of the force can be measured at each point in the field.

Magnetic poles The parts of a magnet from which the magnetic force seems to come. One is the north pole and the other the south pole. Like poles repel each other and opposites attract.

Magnetism The name given to the effects associated with a magnetic field. A magnetic field is produced by magnetic materials, and whenever an electric current flows.

Magnetite A highly magnetic form of the metallic ore iron oxide. Some types of magnetite, known as lodestones, were used as compasses by early sailors and explorers.

Microchip Sometimes shortened simply to chip, a single crystal of a semiconductor that can carry out many purposes in an electronic circuit.

Photon A particle of electromagnetic radiation. Photons are involved in all electromagnetic interactions.

Potash A name given to any of a number of compounds containing the element potassium.

Relay A device in which the current in one electrical circuit is controlled by changing the current in another.

Resinous electricity A type of 'electrical fluid' thought to exist by Charles du Fay. The other type was vitreous electricity.

Resistance A measurement of a material's opposition to the flow of an electric charge through it. The unit is known as the ohm.

Semiconductor A material that has a resistance to electrical current somewhere between that of a conductor and an insulator.

Soda A name given to any of a number of compounds containing sodium.

Solar cell An electric cell that produces electricity from the energy in light.

Solenoid A coil of wire that generates a magnetic field when an electric current is passed through it.

Static electricity The effects produced by electric charges that are at rest and the forces that act between objects that have an electric charge.

Sulphur A chemical element found around volcanoes. At one time this led people to believe that the earth was largely made of sulphur.

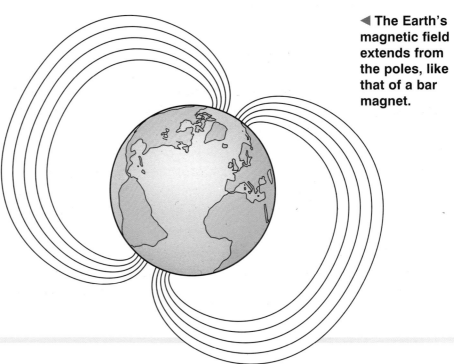

◀ **The Earth's magnetic field extends from the poles, like that of a bar magnet.**

Superconductivity Absence of resistance to an electric current. Occurs in some substances at very low temperatures.

Telegraph Device for sending messages long distances. It uses an electric current which passes along a wire linking a transmitter and receiver.

Transformer A device that transfers electrical energy from one alternating current to another, changing voltage or current in the process.

Transistor Tiny device made from semiconductor material that can act as a switch or amplifier in an electric circuit. Transistors are the basic components of electronic equipment. They were first invented in 1948 by an American team headed by William Shockley.

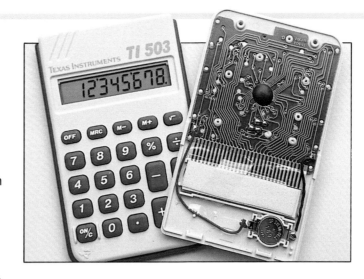

◀ A calculator, with its micro-electronic insides revealed.

Vacuum tube Electronic component, such as a valve or cathode-ray tube, that has had the air extracted from the interior.

Valve Device for controlling the flow of electricity in an electrical circuit. Valves have now been almost universally replaced by transistors, which are smaller, cheaper and more reliable. An exception is some top-class hi-fi equipment. Valves, say the makers, give a sound that seems more musical to the ears of serious hi-fi enthusiasts!

Volt The measure of electrical energy at a particular point. A positive charge moves from a point with high voltage to one with low voltage.

Voltaic pile A form of battery invented by Alessandro Volta, in which cells were joined together, to produce an electric current.

 ## GOING FURTHER

Books There a quite a few books on electricity. Try these for starters:
Eyewitness Science Electricity, Dorling Kindersley.
Discovering Electricity by Neil Ardley, Franklin Watts.
Young Scientist Book of Electricity by Phil Chapman, Usborne.
Magazines General-interest science magazines usually have features on technology with electricity applications. Try the following:
Focus Snippets of information on just about every subject under the sun. Monthly.
Popular Science Packed with great pictures, glossy diagrams and charts. Monthly.

New Scientist A more serious publication that is highly regarded by professionals. Weekly.
Places In Britain, museums with a science bent include the famous *Science Museum* in London, which is particularly strong on historical aspects. The newer, 'hands-on' *Eureka* museum in Halifax is packed with demonstrations and experiments that you can do yourself.
CD-ROM The title list is growing, but there is little that is exclusively to do with electricity. However the Grolier *Encyclopedia* includes many interesting entries, as does the Microsoft *Encarta*. Both feature animations, video clips, text and pictures.

INDEX

alternating current (AC) 28, 29, 44
alternator 30
amber 6, 7, 35, 44
amp, ampere 19, 44
Ampère, André-Marie 19, 42
anode 17, 34

batteries 13, 30, 31, 38, 44
Bell Telephone Laboratories 36

carbon-arc lamp 26
catalyst 44
cathode 17, 38
cathode rays 34, 35, 44
cathode-ray tube 34, 35
cells 30, 31, 44
Claude, Georges 27
compounds 17, 44
conductors 9, 11, 20, 38, 40
Copenhagen University 18
Coulomb, Charles 11, 42
Crookes, William 34

Davy, Humphry 16, 17, 18, 42
De Forest, Lee 36, 43
direct current 44
du Fay, Charles 8, 11, 42

Edison, Thomas Alva 26, 27, 28, 43
Einstein, Albert 38, 43
Einthoven, Willem 14
electric cars 31, 38
electric eel 15
electric motor 22, 23
electrocardiiograph 14
electrodes 17, 44
electrodynamics 5, 19, 45

electroencephalograph 14
electrolysis 16, 17, 45
electrolyte 17, 30, 31, 45
electromagnet 19, 20, 21, 45
electromagnetic induction 24, 29, 45
electromagnetic radiation 38
electromagnetic waves 32, 45
electromagnetism 32
electron 5, 35, 36, 37, 38, 40, 41, 45
electronics 36, 37
electroplating 17
electrostatics 5, 19, 45
elektron 6
energy sources 25

Faraday, Michael 17, 18, 22, 24, 25, 29, 32, 34, 43
Feynman, Richard 41
Fleming, John 36, 43
Franklin, Benjamin 9, 10, 11, 42
fuel cell 31, 45

Galvani, Luigi 12, 14, 42
galvanic electricity 12, 45
galvanometer 24, 45
gas-discharge lamps 27
generator 24, 25, 40, 45
Gilbert, William 7, 8, 42
Gray, Stephen 8

Henry, Joseph 20, 21, 22, 43
Hertz, Heinrich 32, 33, 38, 43

insulators 9, 45
integrated circuits 36, 45

Kilby, Jack 37

Leclanché, Georges 30, 43
Leyden jar 9, 10, 13, 14, 45
light bulbs 26, 27
lightning 10, 11
lightning rods 11
lodestone 6, 7, 46
Lucretius 6
maglev trains 41
magnetism 6, 7, 18, 19, 22, 24, 25, 34, 40, 41, 46
Marconi, Guglielmo 32, 33, 43
Maxwell, James Clerk 32, 38, 43
morse code 21, 33
Morse, Samuel 21

neon lights 27
Nicholson, William 16, 42
Noyce, Robert 37

Oersted, Hans Christian 18, 19, 22, 32, 42
Ohm, George 20, 21, 28, 43
Onnes, Heike Kamerlingh 40, 41

Peregrinus, Peter 7
photoelectric effect 38
photon 38, 46
piezoelectricity 38
Plücker Julius 34
potash, potassium 16, 17, 46
power stations 25, 28, 29

radar 35
radio, radio waves 32, 33, 38

resistance 20, 40, 46
Royal Society 16, 25

Seebeck, Thomas 38
semiconductors 37, 46
shark sensors 15
Shockley, William 36, 43, 47
Siemens company 21
silicon microchip 36, 37, 46
solar cells 38, 39, 46
solenoid 19, 46
Space Shuttle Orbiter 31, 45
static electricity 6, 8, 9, 10, 13, 46
Stoney, George 35
Sturgeon, William 19, 42
superconductors 40, 41, 47
Swan, Joseph 26, 27, 43

telegraphy 20, 21, 33, 47
Tesla, Nikola 28, 43
Thales of Miletus 6, 42
thermoelectricity 38
Thomson, Joseph John 35, 43
transformer 28, 29, 47
transistors 36, 37, 47

valves 36, 37, 47
van Musschenbroek, Pieter 9
Van de Graaff, Robert 9
volt 12, 13, 47
Volta, Alessandro 12, 13, 14, 16, 42, 47
voltaic pile 13, 30, 47
von Kleist, Ewald 9
von Guericke, Otto 8, 19, 42

X-rays 32